SpringerBriefs in Ethics

More information about this series at http://www.springer.com/series/10184

Bashir Jiwani

Clinical Ethics Consultation Toolkit

 Springer

Bashir Jiwani
Fraser Health Ethics Services and Diversity Services
Burnaby, British Columbia, Canada

ISSN 2211-8101   ISSN 2211-811X   (electronic)
SpringerBriefs in Ethics
ISBN 978-3-319-60377-3   ISBN 978-3-319-60379-7   (eBook)
DOI 10.1007/978-3-319-60379-7

Library of Congress Control Number: 2017946845

Printed on acid-free paper

This Springer imprint is published by Springer Nature
The registered company is Springer International Publishing AG
The registered company address is: Gewerbestrasse 11, 6330 Cham, Switzerland

Acknowledgement

Innumerable people have influenced me and the thinking that has led to this book. I am grateful to my professional colleagues at Fraser Health Ethics Services and Diversity Services, my colleagues and mentors in the world of clinical ethics consultation, and my academic mentors for their shaping of my understanding of how this work should be done. I am indebted to my family, especially my wife Nimeera our parents Barkat and Parin Jiwani and Sadrudin and Noorbanu Kassam and my son Rafeeq, for helping me see the work of clinical ethics consultation through the lens of real life. Without Nim, I also wouldn't have had the space to devote to this work - thank you, my love.

I am indebted to Elina Hill for her able editing of various versions of the text. I am immensely grateful for the support and critical feedback of my brilliant and kind colleague Katherine Duthie who has been such a valuable partner in the work of clinical ethics consultation for the past five years. And then there is Susan Rink, my right arm, and sometimes I think my left also, who has facilitated and shepherded the process of writing this book, including providing indispensable editing assistance, who has provided moral support and friendship and without whom there is no way I would have been able to complete this project.

The original version of this book was revised to include the Acknowledgments. An update to this book can be found at DOI 10.1007/978-3-319-60379-7.

Contents

Clinical Ethics Consultation Toolkit

Helping people live with greater integrity.

Bashir Jiwani, PhD

About this Toolkit

This toolkit is designed to assist individuals and teams offering clinical ethics consultation services.

Clinical Ethics Consultation is traditionally offered by those with training in bioethics, who are able to bring an ethics-based decision framework to bear on situations where there is uncertainty or conflict about the best way of moving forward.

Clinical Ethics Consultants combine moral reasoning and analysis with mediation and dispute resolution skills to facilitate shared understandings of issues and values around patient care. Consultants place special focus on patients with a view to developing solutions that enable all consultation participants to live with greater integrity.

This Toolkit supports a method of clinical ethics consultation that is based on the idea that treating people with respect requires treating them kindly and well, listening carefully without judgment, and then engaging perspectives without coercion. Through respectful dialogue, the process involves making patient care decisions guided by the values and beliefs of those to whom a fiduciary duty is owed. Where no such duties exist, it involves deciding based on the values and beliefs of the group, as arrived at collectively.

Reflecting the method of clinical ethics consultation it aids, the Toolkit follows a five stage process: pre-consult, interviews, mid-consult, consult meetings, and post-consult. For rich descriptions of the rationale behind the process, please consult the first sections of the Clinical Ethics Consultation guidebook.

The Toolkit provides a step-by-step guide to this process, with explanations and tips for success, along with worksheets to assist the user to complete each step.

The reader should first review the stages and the overall approach of the consultation process as described in the guidebook.

The Toolkit can be used after a consult intake process has been initiated. Once contact with the requester has been made, the consultant should go to Consult Stage 1 of the Toolkit and follow the process that unfolds.

The consultant may also wish to review the emerging story form in the appendix to anticipate the areas to explore through the consultation process.

FIVE STAGES OF CLINICAL ETHICS CONSULTATION PROCESS

1 Pre-Consult
   - communicate that the consult has been initiated
   - review patient's chart
   - establish consult team and plan
   - communicate consult process plan

2 Interviews
   - prepare for interviews
   - meet with the relevant parties

3 Mid-Consult
   - conduct initial ethics analysis
   - plan consult meeting(s)
   - communicate next steps in the process

4 Consult meeting(s)
   - bring various participants together
   - reconvene as necessary

5 Post-Consult
   - documentation
   - follow up support
   - evaluation
   - identification of systemic issues

Facilitating Conversations

ELEMENTS OF GOOD FACILITATION

For the 5-stage process to be effective, facilitators will have to:

- Create trust within the decision group;
- Clarify and interrogate facts within the group;
- Build a shared understanding of values and a shared commitment to what should matter most;
- Explain each step so participants understand what they're doing and why.

Group facilitation comes with its own skill set, knowledge base, and character traits. Good ethics facilitation requires facilitators to:

1 Demonstrate & help others to demonstrate respect:
- Treat others with unconditional positive regard:
  › "Regardless of what I think of your opinion, I will treat you well."
- Listen empathetically to understand:
  › "I will work hard to open my mind to understand why you think what you think."
  › "I will work hard to open my heart to understand what you are feeling."
- Engage others' ideas:
  › "I will share some of my thinking with the hope of deepening our perspectives and moving to a broader view."

2 Ask good questions:
- Open-ended:
  › Allowing the teller's story to emerge.
- Probing:
  › Exploring the reasons behind perspectives.

3 Reframe responses:
- To demonstrate understanding:
  › So participants are heard and can believe in the process.
  › So the best thinking informs the discussion.
- To neutralize:
  › To make hard-to-hear perspectives more palatable.
  › To enable connections between participants.

- To deepen understanding:
  - › Help tellers hear and reflect on their own starting points of view.

....

4 Distinguish between facts, values, and emotions and help others understand and make these distinctions:
  - Facts:
    - › What is descriptively true in the context.
  - Values:
    - › What is important about the issue.
  - Emotions:
    - › What is in one's heart and body – how one is feeling about the situation.

....

5 Acknowledge facts, values, and emotions without judgment, and help participants work through these individually or collectively:
  - Facts:
    - › This is what you believe is true.
  - Values:
    - › This is what you believe is important.
  - Emotions:
    - › This is what you're feeling about this.

The next few pages offer tips for facilitating the different elements of the process. Additional helpful language and hints can be found in the Tips for Success sections within each step.

DETERMINING FACTS

Facts concern the descriptive backdrop of the story – a description of the world as it is. Who is involved? What is happening? What is the context?

....

Examples of good questions for understanding facts:

Open-ended:

- What is your understanding of the situation?
- How did we arrive at this situation?

- What else do you need to know?
- What else would you like to know?
- If this situation continues unchanged how do you think it will affect:
  - You?
  - Your team?
  - Your other colleagues/the organization?
  - Patients/families?
  - Society at large?

Probing:

- How did you come to this understanding?
- What examples are you thinking of that led you to this concern?
- What makes you think so?
- What evidence are you relying on?

ARTICULATING VALUES

Values are what is important in the story. We can discover our values when we consider what we want more of in the world. Asking good questions can help us determine/understand our values.

Examples of good questions for understanding values:

Open-ended:

- What is important to you as we move forward?
- Why do you prefer this solution – what does it give you that you believe is important?
- What would you like a solution to achieve?
- In our society, this value (e.g., equity) is important – what is your sense of this? What do you think this value means in our context? How important is it?
- What would someone who disagreed with your perspective say is important? How would you respond?

Probing:

- Why is this important?
- Here is a competing value (tell story); how would you balance these two?
- What would have to happen for you to change your mind? What does this tell you about what else matters to you?

SUPPORTING EMOTIONS

How are people feeling? What is going on in their hearts and bodies?

....

Examples of good language for debriefing emotions:

Exploring:

• What is in your heart as you go through this?

• How are you feeling about this?

• You seem very…

• Are you feeling….?

Acknowledging:

• This must be very difficult for you.

• I'm sorry you have to go through this.

It is often difficult for people to discuss how they are feeling. Instead of referring to our emotional state, we often use the word "feeling" to describe what we think about an issue.

Being familiar with different descriptions of emotions can help clarify what a feeling is and open a conversation about emotions. The facilitator may benefit from becoming familiar with these to be better able to help participants describe how they are feeling in or about the situation under discussion.

Consult Stage 1: Pre-Consult

- Helps evaluate the information provided in the intake process.

- Provides a rich source of new information, both about clinical aspects of the patient's situation and about people involved in the patient's life.

- Indicates some of the individuals to be interviewed and suggests lines of discussion for specific parties.

DIRECTIONS

1. Review the intake form and send an initial response.

2. Identify who will be involved in providing the consult.

3. Appoint a consult team leader to ensure all steps in the plan are covered.

   - Complete the Consult Team Form.

4. Review the patient's chart.

5. Populate the Case Summary worksheets with information available so far (from intake process and patient's chart).

6. Identify what information is missing and how it should be obtained.

7. Assign responsibility and develop a plan for conducting interviews.

   - Complete Information Gathering Plan worksheet.

8. Decide how the team will check in during Interview stage, and schedule a meeting to do Mid-Consult analysis and planning.

   - Begin completing Emerging Story worksheet.

9. Review Consult Documentation Form and agree on appropriate times to document consult and revisit documentation plan.

   - Begin completing Consult Documentation Form.

TIPS FOR SUCCESS

- Be secure with the information collected, including interview records – personal and confidential information is often involved.

- Discuss confidentiality, determine what information will be shared and with whom and share this with participants as appropriate.

- Be sure to communicate the consult process with participants so they know what to expect.

- Sitting quietly and reading at a table or desk in a unit (and soaking in the atmosphere) can provide great insight into the culture of the unit.

- If possible, develop your plan together as a consult team while reviewing the chart.

© The Author(s) 2017
B. Jiwani, *Clinical Ethics Consultation Toolkit*, SpringerBriefs in Ethics,
DOI 10.1007/978-3-319-60379-7_1

SAMPLE INTAKE FORM (1 OF 2)

Thank you for your interest in requesting a clinical ethics consultation from Ethics Services.

Please complete this form and send it to our office at:

We will send an acknowledgement of your note upon receipt of your clinical ethics consultation request. This note can be added to the patient's chart, if appropriate. Someone from our office will then be in touch by the next business day to arrange an initial meeting by telephone between yourself and one of our ethics consultants to discuss the case briefly and develop a plan.

REQUESTER INFORMATION

Requester's Name (First, Last):   Date/Time of Request:

Title/Position: Location/Site:

Telephone number(s):   Email address:

Priority:

☐ High (immediate follow-up)

☐ Medium (2-4 weeks)

☐ Low (4-6 weeks)

Requester's Description of the Ethics Case and Concern:

Is part of the difficulty that people see facts in the case differently? (E.g. is there a disagreement about the expected consequences of different options?) If so, what facts are being disputed?

Is part of the difficulty that people have different understandings of what should matter most in the situation? (E.g. is there disagreement about what the care plan should try to achieve?) If so, what is the disagreement?

Type of Assistance Requested (Check all that apply):

☐ Meeting with healthcare team to determine appropriate next steps

☐ Resolving conflict amongst the healthcare team

☐ Meeting with patient/family and team to determine appropriate next steps

☐ Resolving conflict between patient/family/team

☐ Team meeting to debrief past decision (support moral distress amongst the team)

☐ Other (specify): _____

☐ Not sure

SAMPLE INTAKE FORM (2 OF 2)

PATIENT INFORMATION

Patient Name (First, Last):   Age:

Gender: Date Admitted:

Location/Department: Attending Physician's Name:

Was the attending physician notified?   ☐ Yes ☐ No

☐ If no, explain:

Clinical Service:

☐ Critical/Intensive Care   ☐ Home/Community Care
☐ Extended/Long-Term Care   ☐ Mental Health
☐ Medical Subspecialty Care   ☐ Other (please specify): _____
   (please specify): _____

Decision Making Capacity:

☐ Clearly Yes   ☐ Clearly No   ☐ Partial/Fluctuating/Unclear
                                                  (Explain): _____

Is patient affected by one or more morbidities?
Please list all:

PARTIES INVOLVED

FAMILY/LOVED ONES

Name (First, Last):   Relationship to patient: Contact Information:

Name (First, Last):   Relationship to patient: Contact Information:

CARE TEAM

Name (First, Last):   Title/Position: Contact Information:

Name (First, Last):   Title/Position: Contact Information:

How did you hear about our service?

☐ Colleague
☐ Supervisor
☐ Attending an ethics talk or event
☐ Other (specify):

For more information, contact Ethics Services (provide contact info)

Consult Stage 1: Pre-Consult

SAMPLE INITIAL RESPONSE TEMPLATE (to be put in letterhead format)

Ethics Consult for _____

An ethics consult has been requested for the above patient. The consult has been requested by

_____

The purpose of an ethics consultation is to determine the most ethically justified way of proceeding regarding the care of a patient. This process involves: 1) identifying the principle guiding values most relevant to the situation; and 2) determining what direction each of these provides for moving forward. In this case the key question appears to be: What is an appropriate care plan for this patient?

The consult process will first involve reviewing the patient's chart and speaking with various key individuals involved with this situation. Representatives from Ethics Services will be in touch shortly with members of the care team, the patient and the patient's family. Depending upon the circumstances, one or more group meetings may also be arranged. A consult report will then be provided for the patient's chart.

Please do not hesitate to contact me, or members of Ethics Services, should you have any questions or concerns about the consult process or require additional support.

Kind regards,

Signature

Consult Stage 1: Pre-Consult

CONSULT TEAM FORM

Consult Team Member	Email	Telephone		Position in the organization	Consult Lead

INFORMATION GATHERING PLAN FORM

SOURCES TO CONSULT:

- Chart
- Patient
- Family members/loved ones
- Care team members
- Policy and legal experts

Source	Who will meet/review	Time Line
Chart		
Patient		

Family members/loved ones:

Care team members:

Others:

CONSULT DOCUMENTATION FORM

Consult Stage	Documentation			Content						
	Shared with ethics team	Placed on chart	Circulated to all participants	Process update	Key question	Facts summary/explanation	Values summary/explanation	Decision reached/justification	Next steps	Ethics comments / recommendations
After Intake										
After Initial Ethics Analysis and Consult Meeting Planning										
After Consult Meeting										
Interim update										

Consult Stage 2: Interviews

The ethics consultant should try to:

- Understand without judgment:
  - a) interviewee's feelings,
  - b) interviewee's understanding of facts, and
  - c) what is important to her
- Deepen interviewee's perspective of her own story
- Reframe interviewee's story into language that others involved will be able to hear and understand
- Earn the trust of participants
- Model the interview on the process to be used in the consult meeting
- Situate interview within the broader consult process.

DIRECTIONS FOR EACH INTERVIEW

1 Contact participant to set up interview time.

2 Begin Interview Details worksheet.

3 Arrive on time and prepare for conversation.
   - Identify interview space seating arrangement.
   - Confirm the questions you plan to ask.
   - Check in on how you are feeling and make adjustments to become present and put yourself into an active listening mode.

4 Connect with interviewee and undertake the interview, perhaps taking notes as you go.
   - Explain the goals of the consultation service, your role, and what you hope to achieve in the interview.
   - Explain how the interview fits within the broader consultation process.
   - Invite participant to share her perspective.
   - Use open-ended and probing questions, and reframe to understand participant's perspective.

5 Conclude interview.
   - Thank participant for her time.
   - Acknowledge the difficulty of the situation and the consultation process.
   - Remind her of next steps.

6 Review your interview notes (or make them if you haven't yet done so).

7 Decide how the information gained impacts questions to ask in future interviews.

8 Update Emerging Story worksheet with what you have learned.

9 Complete remaining interviews.

10 Share with your colleagues.

TIPS FOR SUCCESS

- Remember, interviews prepare participants for positive discussion, and are crucial for successful consult meetings.
- Use clear and simple language.
- Listen carefully, ask questions, probe opinions for the concerns behind them, suggest more neutral phrasings and expressions.
- Try to move beyond particular positions towards a deeper understanding of facts and values.

© The Author(s) 2017
B. Jiwani, *Clinical Ethics Consultation Toolkit*, SpringerBriefs in Ethics,
DOI 10.1007/978-3-319-60379-7_2

An interview should be seen as a good conversation. Good conversations are transformational - they enrich the perspectives of all involved

- Remember, you are not there to judge, persuade, criticize or to tell your own story.

- At appropriate times, share your understanding of relevant values and ask the participants how their views sit in relation to these social norms.

- Test possible translations of opinions that capture the essence of the participant's perspective in a form more likely to be acceptable to others.

- Clarify to patients and loved ones that you are not a member of the care team, but rather a neutral party – who is a specialist in helping people work through challenging situations toward ethically justified decisions.

- For care team members, emphasize that you are an impartial external party, not there to do the team's bidding, but to help them participate in a broader decision-making environment.

- Be empathetic, thoughtful, and systematic.

- Help care providers distinguish their expert opinions from their personal, evaluative perspectives.

- Attend to the pain the interview might be causing and regularly check in to ensure the participant is comfortable proceeding.

CATEGORIES OF RELEVANT INFORMATION INCLUDE:

Abstract Clinical Facts:
- Disease/illness
- Treatment options
- Epidemiology

Patient Clinical History:
- Diagnosis
- Current treatment
- Treatment options
- Prognosis with various options
- Decision-making capacity

Patient Identity:
- Values & Beliefs
- Preferences
- Agent/Representative
- Personal Directive

Family & Loved Ones:
- Who's involved?
- What's their relationship?
- What are their perspectives?

Professional Caregivers:
- Who's involved?
- What's their assessment?
- What are their perspectives?

System:
- Relevant polices
- Relevant laws
- System dynamics
- Resource issues

Consult Request:
- Reasons for requesting the ethics consultation?
- Who is requesting the ethics consultation?
- What event happened to spur the request?
- Consult objectives?
- Urgency of the issue?
- Consult involvement expected?
- Consult arrangements?

Consult Stage 2: Interviews

SAMPLE INTERVIEW FORM (please duplicate as needed)

How is this person feeling? 1) Ask: How are you feeling? 2) Reframe: So you are feeling_____

What are this person's values? 1) Ask: What is important to you?   2) Probe directly. Why is this important to you?

What does this person believe? 1) Ask: What is your understanding of X? 2) Probe: What makes you think this?

Of the medical facts: Of the patient's identity:

Of the social dynamics of the situation:

Of other relevant features of the case:

Consult Stage 3: Mid-Consult

REFLECT AND PLAN

Ethics analysis and consult planning require focus, discipline, and strategic thinking:

- Focus - avoid distraction, such as talking about issues not directly at hand
- Discipline - systematically review the different types of information in a case
- Strategic thinking - prepare to facilitate complex and emotional conversations about deep subjects in short spaces of time where participants are still working out their views

DIRECTIONS

Meet with your colleagues (as per time set in Stage 1) and in collaboration:

1 Complete Initial Ethics Analysis worksheets:
  a What's the key question?
  b What are the facts:
    i That we all agree on.
    ii That we disagree on.
    iii That we don't know and can find out.
    iv That we don't know and will not know.
  c What's important?
    i To us as consultants?
    ii To the patient?
    iii To family and loved ones?
    iv To the team?
    v To all?
    vi Unique to each?
  d What might possible options be?
  e What communication gaps need to be filled?
  f Who could use emotional support resources?
2 Complete the Consult Meeting Planning Form:
  a What is the purpose?
  b Who should be there?
  c Who will contact which participants?
  d How long should it last?
  e Where will it take place?
  f Who will lead it?
  g What role will each ethics team member play?
  h What will be said in the opening statement?
3 Execute the consult meeting plan

TIPS FOR SUCCESS

- See Ethics Consultation Process in Guidebook for details on the initial ethics analysis steps
- Make time to ensure you get through this stage

© The Author(s) 2017
B. Jiwani, *Clinical Ethics Consultation Toolkit*, SpringerBriefs in Ethics,
DOI 10.1007/978-3-319-60379-7_3

INITIAL ETHICS ANALYSIS: CONFIRMING THE KEY QUESTION(S)

The Key Question the team will focus on:

Questions that need to be addressed:

1

2

3

4

5

6

7

8

Consult Stage 3: Mid-Consult

INITIAL ETHICS ANALYSIS: KEY RELEVANT FACTS

Facts	Details	Is there agreement on this?	
		No	Yes
General clinical information		☐	☐
		☐	☐
		☐	☐
Patient's clinical history	Diagnosis/diagnoses:	☐	☐
	Prognosis:	☐	☐
	Treatment Options:	☐	☐
		☐	☐
		☐	☐
Patient's identity	Patient's values and beliefs:	☐	☐
	Patient's expressed wishes:	☐	☐
	Has the patient identified an agent? If so, who?	☐	☐
		☐	☐
Family & loved ones	Persons impacted/their perspectives:	☐	☐
		☐	☐
		☐	☐
Care team members	Member name/perspective	☐	☐
		☐	☐
		☐	☐
		☐	☐
System		☐	☐
		☐	☐
Ethics consult		☐	☐
		☐	☐

INITIAL ETHICS ANALYSIS: IDENTIFYING AND PRIORITIZING VALUES

Value Theme (Brief name of value)	Value Specification: However we answer the question, it's important that... (The team should add to this list of commonly held values as necessary)	Priority: Important 1	Very 2	Important 3	4	Crucial 5
Respect for patients	We make decisions based on the values and beliefs of the patient.	☐	☐	☐	☐	☐
Patient wellbeing	Our decision advances the wellbeing of the patient, from her perspective.	☐	☐	☐	☐	☐
Fairness	We do not penalize patients for attributes that are beyond their control.	☐	☐	☐	☐	☐
Community wellbeing	We do not cause harm to others involved in the situation.	☐	☐	☐	☐	☐
Respect for family	The decision reflects the values and beliefs of family members.	☐	☐	☐	☐	☐
Trust	We build trust with the patient and their loved ones.	☐	☐	☐	☐	☐
Respect for patients	We involve family members and loved ones as as the patient wants or would want.	☐	☐	☐	☐	☐
Care provider integrity	We honour the integrity of care team members.	☐	☐	☐	☐	☐
Care for the vulnerable	We pay particular attention to those who are vulnerable.	☐	☐	☐	☐	☐
Patient wellbeing	We do not cause harm to the patient.	☐	☐	☐	☐	☐
		☐	☐	☐	☐	☐
		☐	☐	☐	☐	☐
		☐	☐	☐	☐	☐
		☐	☐	☐	☐	☐
		☐	☐	☐	☐	☐
		☐	☐	☐	☐	☐
		☐	☐	☐	☐	☐
		☐	☐	☐	☐	☐
		☐	☐	☐	☐	☐

INITIAL ETHICS ANALYSIS: ANTICIPATING AND EXAMINING OPTIONS

→ Possible ways of answering the key question (from INITIAL ETHICS ANALYSIS: CONFIRMING THE KEY QUESTION(S) worksheet) include: ↓ How well does the possible option in this column allow us to (list value statements in priority order):	Possibility 1:	Possibility 2:	Possibility 3:	Possibility 4:	Possibility 5:
1.					
2.					
3.					
4.					
5.					
6.					
7.					

CONSULT MEETING PLANNING FORM - #1 OF 2

Meeting Details

Where the meeting will be held

Who will do the booking

Start time

Duration

Participants	Who will be invited	Who will invite them
Patient		
Family & Loved Ones		
Care Team members		
Others		
Ethics Team members		

Meeting objectives for Ethics Team

Ideal seating arrangements (sketch)

Opening statement

Content to be covered

Introduction to ethics services

Overview of the ethics process

Discussion guidelines

Meeting length

Confidentiality

Confirmation of commitment to proceed

Authority to settle

Quick introductions

CONSULT MEETING PLANNING FORM - #2 OF 2

Meeting roles

Role	Team member responsible
Meeting chair	
Timekeeper	
Scribe/Notetaker	

Facilitation responsibilities

Task	Team member responsible
Opening remarks	
Establishing issues and the key question	
Getting a shared understanding of the facts	
Getting a shared understanding of values that matter to each participant	
Getting a shared understanding of what should matter most	
Brainstorming options	
Evaluating options	
Detailing the best options	
Next steps	
Closing	

Next Steps

To Do	Team Member responsible	Timeline

Consult Stage 4: Consult Meeting(s)

WHAT TO COVER IN THE OPENING STATEMENT

- Overview of ethics services, including consultation
- Relationship between consult team and health care team
- Overview of ethics process
- Introduction of consult team members
- Discussion guidelines
- Meeting length
- Confidentiality
- Authority to settle
- Confirmation of commitment to proceed under these terms

DIRECTIONS:

After the Mid-Consult work is complete, this step is about the group meeting with the relevant participants involved in the situation.

1. Psychologically prepare on the day of the meeting.
2. Arrive at the venue a few minutes early to meet with your consultant colleagues.
3. Set up the room, move chairs, prepare flip charts and whatever else you have discussed in your mid-consult planning.
4. Confirm with your colleagues who will be playing what roles in the meeting.
5. Welcome participants.
6. When all participants have arrived, offer a collective welcome, begin the meeting, and go through the various steps in the process:
   - Offer opening remarks.
   - Establish issues and the key question.
   - Get a shared understanding of the facts.
   - Get a shared understanding of each participant's values.
   - Get a shared understanding of the relative priority of the values identified.
   - Brainstorm options.
   - Evaluate options against values.
   - Detail the best option.
   - Determine next steps.
7. Adjourn the meeting.
8. Meet with your colleagues to debrief and plan interim measures.

TIPS FOR SUCCESS

- Divide responsibility for facilitation amongst the consult team.
- Highlight the importance of focus and discipline in the meeting.
- Own the process - take charge of how the meeting is run.
- If you get stuck (which always happens), don't panic! Just acknowledge you're stuck and find your way back into the process.
- Don't own the problem. Remember, you are there only because you have tools that can help participants arrive at an ethically justified solution.

© The Author(s) 2017
B. Jiwani, *Clinical Ethics Consultation Toolkit*, SpringerBriefs in Ethics,
DOI 10.1007/978-3-319-60379-7_4

Consult Stage 4: Consult Meeting(s)

CONSULT MEETING DATA CAPTURE FORM

Issues/questions
identified:

Key question
everyone agreed
to focus on:

Agreed
upon facts/
understandings:

Consult Stage 4: Consult Meeting(s)

Disagreed upon
understandings:

Questions about
the facts that still
require answers:

Values everyone
agreed upon:

Values
participants
disagreed about:

Possible solutions
identified:

Consult Stage 4: Consult Meeting(s)

Analysis of possible solutions:	Option:	Option:	Option:	Option:
	Analysis:	Analysis:	Analysis:	Analysis:

Decisions reached:

Justification for decision:

Next steps:

Consult Stage 5: Post-Consult

ELEMENTS OF POST-
CONSULTATION

- Documentation
- Follow up support
- Evaluation
- Identification of systemic issues

DIRECTIONS:

- Complete consult documentation.
- Set a time to meet with consult team colleagues to formally review the meeting and evaluate the consult process and experience.
- Undertake the broader consult-specific evaluation plan for the consult service.
- Complete the team support worksheet and develop an action plan for following through.
- With the appropriate members of the consult service team, complete the Systemic Analysis Form.

TIPS FOR SUCCESS

- Doing post-consultation well requires systemic strategies for each of the dimensions.
- For external evaluation, a method and tools need to be put in place that reflect the questions on page 31.

© The Author(s) 2017
B. Jiwani, *Clinical Ethics Consultation Toolkit*, SpringerBriefs in Ethics,
DOI 10.1007/978-3-319-60379-7_5

Consult Stage 5: Post-Consult

GUIDING QUESTIONS FOR CONSULT PARTICIPANTS

Evaluation Question	Responses Collected	Response Plan (if any)
Was the tone of the consult team members genuinely respectful during interviews?		
Did ethics team members appear to be comfortable in their role?		
Were all participants' concerns made explicit in the discussion?		
Did ethics team members explain their role well?		
Were ethics team members transparent about the process they were using?		
Did ethics team members ask questions that helped clarify what information was known, contested, and unknown?		
Did ethics team members ask questions that helped you get a deeper understanding of your own values?		
Did ethics team members bring to light other values that you might want to consider?		
Did ethics team members answer your questions?		
Did you feel heard throughout this process?		
Was the decision arrived at and understood by all involved? If not, please explain.		
Retrospectively is the decision viewed as the best one possible?		
Were the conclusions that were drawn temporary, tentative, or definite?		
Do you have a plan for moving forward as a result of the discussion?		
How has this process educated health care providers to help them manage future issues?		
Would you use ethics consult services for support in the future?		
Do you feel better prepared to deal with a similar issue in the future?		

TEAM SUPPORT FOLLOW-UP FORM

Question	Action Items
Do the care team members involved have a shared understanding of the basic beliefs implicit in their practice?	
Are care team members comfortable raising hard questions?	
Is there a mechanism in place for having conversations about difficult or uncomfortable issues?	
What work has the care team done in developing their ethics capacity and what are the next steps they should take?	
Would the care team like more ethics education?	
What types of educational supports might be useful for them?	
When will we check in with the care team again to see how they're doing?	
Who will do this?	

Consult Stage 5: Post-Consult

SYSTEMIC ANALYSIS FORM

Questions   Action Items

What systemic structure or events that happened earlier, in the patient's experience or the team's experience, gave rise to the issue?

What could have been done upstream to prevent the issue from arising?

Can any recommendations be made to effect upstream change?

Who should make these recommendations?

Who should these recommendations be made to?

Appendix: Emerging Story Form

EMERGING STORY FORM (1 OF 8)

General Clinical Information	If we know this for sure, based on what evidence?	If we don't know, can we find out?	If so, who will do this research?
Illness/disease name:			
Relevant details about the illness/disease:			
Possible treatment options for the illness/disease:			
Possible symptom management options for the illness/disease:			

Additional illness/disease name:	If we know this for sure, based on what evidence?	If we don't know, can we find out?	If so, who will do this research?
Relevant details about the illness/disease:			
Possible treatment options for the illness/disease:			
Possible symptom management options for the illness/disease:			

EMERGING STORY FORM (2 OF 8)

The patient's clinical situation			If we know this for sure, based on what evidence?	If we don't know, can we find out?	If so, who will do this research?
Diagnosis & current treatment	Condition	Current treatments			
	Condition	Current treatments			
	Condition	Current treatments			
	Condition	Current treatments			
	Condition	Current treatments			
Patient's current ability to understand their own medical condition					
Patient's current ability to think about and formulate their values and beliefs					
Condition for which treatment is disagreed upon					
Possible treatment options for condition	Treatment option	Prognosis with this option			
	Treatment option	Prognosis with this option			
	Treatment option	Prognosis with this option			
	Treatment option	Prognosis with this option			
	Treatment option	Prognosis with this option			
Hospital / Health care facility history					
Other details relevant to patient's clinical background					

EMERGING STORY FORM (3 OF 8)

The patient's identity - who they are as a person		If we know this for sure, based on what evidence?	If we don't know, can we find out?	If so, who will do this research?
Patient's feelings	What emotions are they experiencing?			
Patient's understanding of clinical facts	What does the patient understand about her/his health condition, treatment options, and prognosis?			
Patient's expressed preferences	What does the patient say she/he wants – in terms of treatment, placement, activities, and relationships?			
Patient's important beliefs	What does the patient believe about the purpose of life, what happens after death, the role of suffering in life, etc.?			
Patient's important beliefs about their health care	What special significance does the patient attach to specific treatment modalities? What concerns do they have about specific parts of their treatment plan?			
Patient's values	What makes their life meaningful?			
Important relationships	Who is the patient close to and whose support does she/he want?			
Important activities	What activities does the patient find worthwhile? Which activities make the patient's life meaningful?			
Substitute decision-making	Has the patient identified someone they would like to make decisions on their behalf if they are unable to do so?			
Substitute decision-making	What is important to the patient about who makes decisions on her behalf, if she is not able to do so? How would she want these people to work together, and what would she want them to be mindful of?			
Resolving disagreement	How does the patient want disagreement to be resolved among decision-makers? Should one person's perspective count for more than others'? Must they agree?			

EMERGING STORY FORM (4 OF 8)

Family members' and loved ones' perspectives			If we know this for sure, based on what evidence?	If we don't know, can we find out?	If so, who will do this research?
Name	Relation to the patient	Understanding of relevant facts.			
		Understanding of patient's values.			
		What she/he wants for the patient – their hopes for the patient.			
		What she/he wants for themselves in the situation – their own personal hopes.			
Name	Relation to the patient	Understanding of relevant facts.			
		Understanding of patient's values.			
		What she/he wants for the patient – their hopes for the patient.			
		What she/he wants for themselves in the situation – their own personal hopes.			

Appendix: Emerging Story Form

Family members' and loved ones' perspectives			If we know this for sure, based on what evidence?	If we don't know, can we find out?	If so, who will do this research?
Name	Relation to the patient	Understanding of relevant facts.			
		Understanding of patient's values.			
		What she/he wants for the patient – their hopes for the patient.			
		What he/she wants for themselves in the situation – their own personal hopes.			
Name	Relation to the patient	Understanding of relevant facts.			
		Understanding of patient's values.			
		What she/he wants for the patient – their hopes for the patient.			
		What she/he wants for themselves in the situation – their own personal hopes.			

Care team – individual perspectives			If we know this for sure, based on what evidence?	If we don't know, can we find out?	If so, who will do this research?
Name	Position/role	Understanding of relevant facts.			
		Understanding of patient's values.			
		What she/he wants for the patient – their hopes for the patient.			
		What she/he wants for themselves in the situation – their own personal hopes.			

EMERGING STORY FORM (6 OF 8)

Care team – individual perspectives			If we know this for sure, based on what evidence?	If we don't know, can we find out?	If so, who will do this research?
Name	Position/role	Understanding of relevant facts.			
		Understanding of patient's values.			
		What she/he wants for the patient – their hopes for the patient.			
		What she/he wants for themselves in the situation – their own personal hopes.			
Name	Position/role	Understanding of relevant facts.			
		Understanding of patient's values.			
		What she/he wants for the patient – their hopes for the patient.			
		What she/he wants for themselves in the situation – their own personal hopes.			
Name	Position/role	Understanding of relevant facts.			
		Understanding of patient's values.			
		What she/he wants for the patient – their hopes for the patient.			
		What she/he wants for themselves in the situation – their own personal hopes.			

Appendix: Emerging Story Form

Care team – individual perspectives			If we know this for sure, based on what evidence?	If we don't know, can we find out?	If so, who will do this research?
Name	Position/role	Understanding of relevant facts.			
		Understanding of patient's values.			
		What she/he wants for the patient – their hopes for the patient.			
		What she/he wants for themselves in the situation – their own personal hopes.			

Care teams, services, or programs involved in the patient's care	If we know this for sure, based on what evidence?	If we don't know, can we find out?	If so, who will do this research?
Name of team/service/program involved:			
Do members of this care team have a shared sense of the facts?			
Do members of this care team have a shared sense of what matters most in the situation?			
Is there a sense of comfort and safety amongst the team such that any team member can raise questions or concerns?			
Is there a regular opportunity for team members to come together to discuss issues of concern that are raised?			
Are there important team dynamic or context issues that are impacting the situation?			

EMERGING STORY FORM (8 OF 8)

System Issues: Inter-team dynamics, policies, laws	If we know this for sure, based on what evidence?	If we don't know, can we find out?	If so, who will do this research?
If there is more than one team involved in the patient's care, are they working in an integrated and mutually supportive way?			
What broader issues at the facility are impacting different teams' abilities to work together?			
What organizational policies are relevant to the situation?			
What laws or legal concerns are relevant to the situation?			

Consult-request	If we know this for sure, based on what evidence?	If we don't know, can we find out?	If so, who will do this research?
What particular event happen to spur the consult request?			
What made the team believe an ethics consult would be appropriate?			
What are the expectations of the team regarding the consult service?